HURRICANE HUNTERS AND TORNADO CHASERS

Life in the Eye of the Storm

Lois Sakany

To my son, Isaiah

Published in 2003 by The Rosen Publishing Group, Inc.
29 East 21st Street, New York, NY 10010

Copyright © 2003 by The Rosen Publishing Group, Inc.

First Edition

All rights reserved. No part of this book may be reproduced in any form without permission in writing from the publisher, except by a reviewer.

Library of Congress Cataloging-in-Publication Data

Sakany, Lois.
Hurricane hunters and tornado chasers / Lois Sakany.
p. cm. — (Extreme careers)
Includes bibliographical references and index.
Summary: Describes the destructive forces behind severe weather systems such as tornadoes and hurricanes and the experiences of people interested in researching and tracking these storms.
ISBN 0-8239-3634-1
1. Hurricanes—Juvenile literature. 2. Tornadoes—Juvenile literature. 3. Meteorologists—Juvenile literature. [1. Hurricanes. 2. Tornadoes. 3. Meteorologists.] I. Title. II. Series.
QC944.2 .S24 2002
551.55'2—dc21

2001007019

Manufactured in the United States of America

551.552
SAK
2003

Contents

Life on the Run

Racing past uprooted trees and downed power lines, most storm chasers plow fast into teeming rain and hail to catch a glimpse of a major storm in action. There is little time to react, and most rely upon years of field work, tracking storms and their behavior by gathering plenty of concrete data such as current air temperature, pressure, and wind speed. Their efforts seem dramatic to many, but so are the results of witnessing a tornado or hurricane in action.

Most storm chasers almost never recall sunny, mild days with cloudless skies, but choose instead to remember their closest calls with disaster. They prefer living life on the edge of extreme, but natural, phenomena. Storm chasers, the subject of this book, are a unique and devoted group of people who track severe weather patterns such as tornadoes and hurricanes for pleasure, research, and, occasionally, profit. When everyone else has run for cover, you can almost bet that there are relentless storm hunters on the loose, hoping to reveal the scope and scale of a major weather event as it happens.

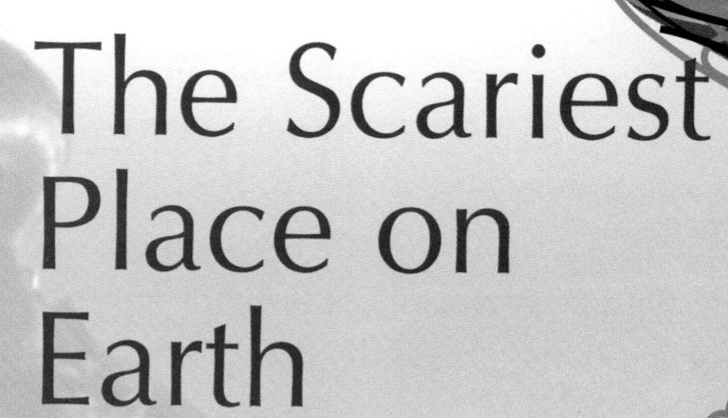

The Scariest Place on Earth

1

Intensely powerful and completely unstoppable, tornadoes and hurricanes—two of the most extreme forms of weather on the planet—descend upon cities large and small each year. These are storms that wreak complete havoc, sometimes taking lives and often leaving millions of dollars worth of property damage in their wake. Listen to the stories of anyone who has crossed paths with a tornado or hurricane, and they just might describe it as the scariest thing on Earth.

When residents of Oklahoma and Kansas witnessed a group of tornadoes in 1999, it went quite beyond what people who live in "Tornado Alley" experience in a normal storm season. In fact, to Oklahoma governor Frank Keating, "The extent of the devastation was unprecedented." Although quite familiar with the usual turmoil associated with tornadoes—terrific wind speeds and beating rain and hail—this destruction was worse. Part of the reason was because more than one tornado had touched down over the state line. As reported by survivors and other witnesses, the destruction was the result of a swarm of tornadoes—sometimes called twisters—that touched down between May 3 and 4 of that year. Later reports, confirmed by the National Oceanic and Atmospheric Administration (NOAA),

Tornadoes such as this one are common in Oklahoma and Kansas and are responsible for the deaths of more than forty people in the United States each year.

estimated that forty-five twisters had hit the area, some exhibiting wind speeds in excess of 207 miles per hour (mph).

Imagine, for a minute, what it was like to witness one of those tornadoes. Touching down on a spring evening, they tore through parts of Oklahoma and Kansas for more than an hour, leveling entire neighborhoods and killing forty-nine people. Fierce winds ripped through rush-hour traffic along a major highway and transformed it into a junkyard as dozens of cars and trucks, each weighing several tons, were tossed around like small toys. Afterward, witnesses compared the city to a battlefield. They felt as if a hydrogen bomb had hit them.

Adding to the risk, other areas in the storm front were pelted by softball-sized chunks of hail. Entire neighborhoods were destroyed, and an estimated 1,000 homes were damaged in Oklahoma City alone. Covering nineteen square miles of land, estimates place the cost of the damage at just over $1 billion.

Almost everyone has seen Hollywood images of twisters and violent storms. For instance, if you've seen either *The Wizard of Oz* or *Twister*, you might think that tornadoes are just as powerful or even more powerful than hurricanes. In reality, however, hurricanes sometimes last for several days and actually cause more damage on a yearly basis than tornadoes do. Hurricane Andrew, a storm that pounded Florida's coast in August 1992, was by far the most expensive natural disaster in history. More than sixty people were killed, and approximately two million people were evacuated from their homes. The estimated cost of damages was a staggering $20 billion.

Scouting a Storm

As all experienced storm chasers know, there are distinctive signs to watch for when scouting the skies for a potential twister in the making. The most common signs include:

- An odd color in the sky, usually greenish in tone
- A period of quiet just after a thunderstorm
- Fast-moving clouds in one area, especially in a rotating pattern
- A loud roaring sound
- Branches or leaves appearing to move upward in the wind
- Hail

Life in a Whirlwind

As frightening as tornadoes and hurricanes might be to most people, there are individuals who find tornadoes and hurricanes so fascinating that they actually seek them out. Known as storm chasers, their aim is to get as close as possible to the heart of a storm and, in the case of hurricanes, fly right inside!

It may sound crazy that someone might actually chase after, rather than run from, a tornado or hurricane,

Storm chasers get as close to storms as possible to document these rare phenomena, often following the active path of a tornado or hurricane.

but people who witness these weather events see them as a fascinating glimpse into natural power. If you find that you follow news reports of extreme weather patterns, hoping for the chance to witness a tornado or hurricane up close, then you may have what it takes to be a storm chaser. You may find yourself agreeing with most chasers who, even though they know severe storms are deadly, also see them as intriguing and even strangely beautiful.

That is how Martin Lisius feels. An experienced storm chaser and the president of Tempest Tours, a business that takes groups of people on tornado chasing expeditions, he compares the windstorms to precious jewels. "Significant tornadoes require a delicate balance of conditions coming together at the same time and the same place. The occurrence of very severe tornadoes is actually very rare, so for me, seeing one is almost like seeing a diamond."

An Age-Old Mystery

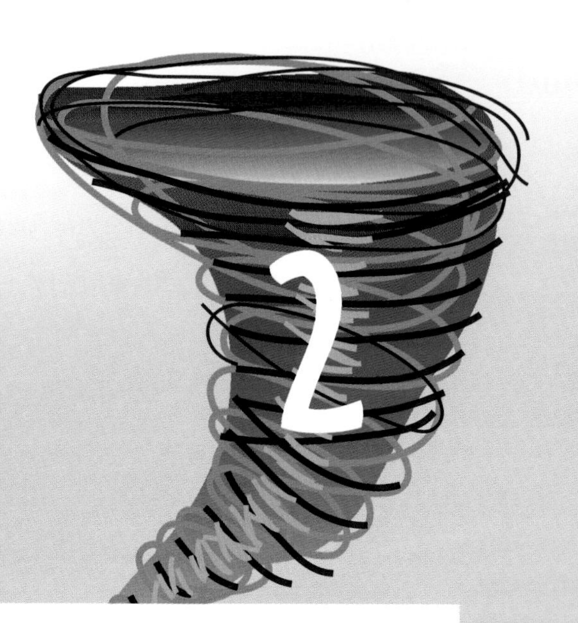

The fascination people have with storms, and for that matter, weather, is not new. The well-known Greek philosopher Aristotle, who was born in 384 BC, wrote a book called *Metereorologica*. This book was the first one to study weather patterns, which is now known as meteorology. Looking back, some of Aristotle's theories about storms may now seem primitive, but the important thing to remember is that he did not have the technology that we have today.

Despite attempts by ancient scientists to provide answers, the tremendous power of and seemingly random patterns of storms convinced many that tornadoes and hurricanes were acts of God. During the Middle Ages in England, people believed that ringing church bells during storms warded off lightning. The practice continued even though one historical observer counted more than 100 bell ringers who were killed in three decades.

America's First Storm Chaser

Ben Franklin, an American statesman and diplomat, is famous for the studies he conducted on lightning. He is also credited with the invention of the

Benjamin Franklin (1706–1790), who declared that a storm's course could be plotted, was among the first and most well-known storm chasers.

lightning rod—a metal rod placed high on a building that conducts lightning safely to the ground. This device continues to save thousands of homes and lives every year. Franklin may also qualify as the first storm chaser of record. In 1775, while riding horses with friends, he wrote about "coming across a small whirlwind beginning in the road and showing itself by the dust it raised and contained." Did Franklin turn his horse and run for his life? Of course he didn't. Setting the standard for generations of modern storm chasers, he moved closer to it and even tried to "break it" by striking his riding whip through it!

The Science of Weather

Though we have been studying severe weather for a long time, our ability to predict storms is a new science. Before the computers, radar, and satellites that we now take for granted, weather forecasters largely depended upon word-of-mouth reports on approaching storms.

In 1938, for example, forecasters had predicted that a hurricane would hit the Florida coast until they received ship reports that the storm was moving up and away from the shoreline. Floridians breathed a sigh of relief, unboarded their windows, and went about their daily business. And while the storm did indeed pass by the state that day, under the cover of darkness it barreled up the U.S. coast moving north, but not away from the shoreline.

The next morning, when beachfront residents of Long Island, New York, looked out at the ocean, they were surprised to see a heavy fog quickly rolling in their direction. To their shock and horror, the "fog" was actually

A car dangles over a broken bridge in West Hampton, New York, a casualty of a hurricane that claimed hundreds of lives on the East Coast in 1938.

a twenty-foot wave of hurricane-driven water. Known as the Long Island Express, the storm killed hundreds of people on and near the beaches of New York and Rhode Island who had no idea a storm was coming. It also flooded downtown Providence and flattened millions of trees in northern New England.

Nowadays, forecasters are much more prepared to predict severe storms. Still, there are many scientific mysteries that remain concerning tornadoes and hurricanes. Daphne Zaras, whose job description as research meteorologist with the National Severe Storms Laboratory, part of NOAA, includes chasing tornadoes, says, "Right now we know a lot about the types of storms that tornadoes come from, but we still don't know exactly when and why tornadoes form."

While both tornadoes and hurricanes are classified as extreme storms and share similar properties, they are two distinct phenomena. If you are thinking that you might want to chase severe storms as a profession, you will probably find that most chasers focus on either tornadoes or hurricanes. Also, scientists who both study and chase severe storms work in research laboratories that specifically focus on either tornadoes or hurricanes.

3 And Toto, Too!

Whenever and wherever conditions are right, tornadoes are possible. Scientists now know that when warm air rises, condenses into clouds, and becomes water, the stage is set for tornado activity. This water, or condensation, stores a tremendous amount of energy, just like a thunderstorm does. When the warm air and energy meet colder air on the surface of the ground, they sometimes produce a vortex, the circling winds commonly found in twisters.

In North America, tornadoes are most common in the central plains of the United States, east of the Rocky Mountains and west of the Appalachian Mountains. They occur mostly during the spring and summer months; the tornado season comes early in the South and later in the North. Scientists also know that tornadoes usually occur during the late afternoon and early evening. However, they have been known to occur in every state in the United States, on any day of the year, and at any hour. All the continents of the world are also affected.

Blowing in the Wind

Tornadoes are formed from a special type of thunderstorm called a super cell. What differentiates super cells from other storms is that they rotate. They can also be huge—as many as fifteen miles wide and sometimes more than 50,000 feet tall. All that rotation creates an abundant amount of energy. Tornadoes—rotating, funnel-shaped clouds—come from a release of that same energy. For example, a tornado with wind speeds of 200 mph is roughly equivalent to the same power of a large nuclear reactor—the type of energy force used to electrify some major U.S. cities!

As powerful as they are, tornadoes account for only a very small percentage of the energy in a thunderstorm. What makes them dangerous is that their energy is concentrated in a small area. Not every tornado is the same, and one of the questions to which scientists are seeking answers is how portions of a thunderstorm's energy sometimes focus into something as small as a tornado.

Other than size, another difference between tornadoes is strength. In order to measure this unbelievable power, scientists use the Fujita-Pearson Tornado Scale, which rates a tornado from zero to five based

SIMMS LIBRARY
ALBUQUERQUE ACADEMY

Meteorologist Tetsuya Fujita, shown here with his tornado simulator, developed the international standard for measuring tornado intensity.

upon the amount of damage it unleashes. For instance, a tornado that measures F-0 causes only small amounts of damage, such as cracks or breaks in chimneys and tree branches. On the severe end of the scale is the F-5, a tornado that is capable of lifting homes from their foundations and tossing cars and other heavy objects with ease.

Martin Lisius, who has tracked hundreds of storms, has seen the effects firsthand of many F-5 tornadoes. "I saw a tornado that ripped asphalt off a paved road, right down to the dirt. That same year, there was a tornado that went through some Texas towns and it picked up a railroad car and dragged it across the pavement, which was left with a huge gouge," he said. Later that year, Lisius missed another F-5 in Oklahoma, but he did arrive in time to see the damage. "The houses were stripped down to the foundation—there was nothing left, no carpet, no plumbing, nothing but sheer bolts."

One thing you might not guess about tornadoes is that the sound they make is as terrifying as their appearance. People often compare the noise to fast-moving freight trains, only much louder. And, because of their ability to form and move quickly,

Flying Feathers: Myth or Reality?

Whenever a tornado strikes, stories of what it has left in its path soon follow. In twister territory, tornadoes have an established reputation for an appetite for destruction, as well as what appears to be prankish behavior. For example, an entire house might be leveled with the exception of a hutch full of fine china left exactly in place without a mark. Sometimes, it is hard to distinguish real-life occurrences from tall tales. There are stories, for example, of iron jugs blown inside out or, in the same vein, the story of a rooster that was blown into a jug with only its head sticking out of the container's neck.

One strange reoccurring story is that of chickens and other birds being found alive, but stripped of their feathers, after a tornado has occurred. Theories abound as to how this phenomenon develops. At one time it was thought that the feathers explode off the bird in a tornado's low pressure. Scientists, however, questioned why only the feathers blew up. After all, they theorized, if a tornado has the ability to blow those to smithereens, wouldn't the chicken be reduced to parts as well? Current theories suggest that the most likely explanation for the defeathering of chickens is the protective response called flight molt. Chickens are not stripped clean, but lose a large percentage of their feathers under stress in this molting process. In a predator-chicken chase situation, flight molt would give the predator a mouth full of feathers instead of fowl!

An image of a tornado's wrath from the film *Twister*, which portrayed the exploits of tornado chasers

sometimes that noise is the only warning people have of a tornado's arrival. In the movie *Twister* the sound designers tried re-creating the noise of a tornado using everything from roller coasters to roaring lions, but in the end, settled on a slightly altered recording of camels!

4

Eyeing the Storm

The word "hurricane" originated from a Caribbean Indian word that, when translated, means, "evil spirit and big wind." Because it combines both the destructive forces of wind and a storm surge (a wall of water as much as twenty feet higher than normal high tide), there is little chance of escape for anyone who is caught near the coast or on a small island during a hurricane. Even if you don't drown, it is difficult to find shelter. Not only are buildings in danger of crumbling, but flying glass, wood, and metal fill the air.

Just like tornadoes, hurricanes are seasonal. They are generally a summer phenomenon, but the length of the hurricane season varies depending on geographic location. For example, the Atlantic hurricane season officially starts on June 1 and ends November 30, but most tropical storms and hurricanes occur between August 15 and October 15. In some parts of the world, hurricanes are known as typhoons.

Hurricanes form in two distinct phases. The first phase is called the genesis stage. Some storms fizzle out in this first phase, but those that don't enter the intensification stage. Just as the description implies, storms in the second phase, like a fighter in training, get bigger and stronger. One essential ingredient to a hurricane's development is warm water. Without water that is at least 80° Fahrenheit, the storm dies. Also, that warm water must be deep. Hurricanes churn the ocean at great depths, and if all that mixing brings cold water to the surface, the hurricane will have difficulty intensifying.

The Name Game

When hurricanes reach a certain size they are assigned names, a practice that started during World War II when forecasters began informally naming tropical storms after their girlfriends and wives. In the late

Hugh Willoughby directs the Hurricane Research Division of the National Oceanic and Atmospheric Administration in Miami.

1940s, the system of assigning female names became official, and in the 1970s, male names were added to the hurricane name list. Hurricane names are recycled every six years, although if a hurricane does a great deal of damage or for some other reason is especially memorable, its name is retired. "Andrew," for instance, will never be used in the future to name another hurricane.

What is probably most incredible about hurricanes, besides the damage they can inflict, is their appearance.

An infrared image of Hurricane Andrew, which ravaged the Caribbean and southern Florida in August 1992.

Simply put, hurricanes are huge—so large that forecasters can easily track their movements by satellite. On average, the width of a hurricane is 500 miles, though they can be much larger. Bigger, however, doesn't always mean more destructive. Some very small hurricanes—sometimes called midgets—can do as much or more damage than much larger storms.

If you look at any satellite photograph of a hurricane, the first thing you might notice is that it seems to have a hole in its center. This hole or depression is called the eye, and it can vary from 50 to 100 miles wide. The skies surrounding the center of the hurricane may be boiling with chaos, but inside the eye, all is eerily calm. Many times, in fact, people have mistakenly left their shelters thinking that a storm has passed, when in reality they were experiencing the storm's midpoint—its eye—and the worst was yet to come.

Weather forecasters use the Saffir-Simpson Hurricane Scale, which ranges from one to five, to measure a hurricane's intensity. A category one hurricane may damage unanchored mobile homes, shrubbery, and trees. It can also cause coastal flooding and minor pier damage. Category three or higher hurricanes are considered major storms. Hurricanes like Andrew are category five storms that not only destroy buildings, but also cause major flooding. If you live near the coast and forecasters predict that a category five hurricane is headed in your direction, you will most likely be evacuated from your home.

When asked to name a particularly destructive hurricane, most Americans would probably first mention

Storm Troopers

As a member of the 53rd Weather Reconnaissance Squadron, (a.k.a. the Hurricane Hunters) it is Major Richard Henning's job to actually fly into hurricanes. As a flight meteorologist, it is his job to observe storms and incoming data. Major Henning explains what is it like to be in the middle of a storm as strong and unpredictable as a hurricane:

> *When we take off, it might be a regular, sunny day, but as we approach the hurricane, we start to fly through bands of showers. Then as we get closer to the eyewall [a thick wall of clouds that surrounds the eye], it gets a lot darker and the rain starts hitting the plane harder to the point where it starts seeping into the plane through the windows. In a severe hurricane, it's like an amusement park ride going up and down, only a lot more jarring. At its worst, if you're not strapped in, you'll be bounced out of your seat. In the eye, the turbulence stops, the winds drop from 100 mph to almost zero, and you suddenly break out into a cloud-free area. As we fly out of the center, the bad weather starts up again. Once we're nearly out, we'll crisscross back through the center four or five more times. I've been doing this for six years, and so far I've never been scared. For me, it's a really interesting job that's sometimes exciting.*

Hurricane Andrew. For residents of Central America, however, the mere mention of the name Mitch is bound to conjure up bad memories. In 1998, Hurricane Mitch dumped seventy-five inches of rainfall in the mountainous regions of Central America, resulting in floods and mud slides that destroyed thousands of homes and killed an estimated 9,000 to 18,000 people. Not since the great hurricane of 1780,

This wreckage, formerly a trailer park in Homestead, Florida, was caused in August 1992. Hurricane Andrew's winds, in excess of 164 mph, caused severe damage to the state.

which killed 22,000 people in the eastern Caribbean, was there a more deadly hurricane.

One important thing to remember about hurricanes is that they are a normal part of nature. As merciless as they may seem, many scientists believe that some of the destruction they do is actually beneficial to the environment. They provide rain that can relieve drought-stricken coasts, while flooding has been shown to flush out bays of pollutants and restore new vitality. Hurricanes have also been known to correct human error. For example, after a non-native pine tree was planted on Key Biscayne (an island off the southern coast of Florida), it not only flourished, but also pushed out native plants. Unlike the local flora, however, it was unable to withstand the force of a hurricane and was eliminated when Hurricane Andrew hit in the early 1990s.

5 Chasing a Dream

If, like research meteorologist Daphne Zaras and other storm chasers, you find the idea of getting a glimpse of severe storms in action exciting, you might want to consider a career that includes chasing tornadoes or hurricanes. One thing to remember is that there is no such job as a full-time salaried storm chaser. All storms are seasonal, and as a result, no person chases tornadoes or hurricanes on a full-time basis. Zaras spends only six weeks out of her whole year in the field searching for tornadoes, and Major Henning, who flies into hurricanes, does so only during a sixty-day period each year. There are a few photographers

A History of Severe Weather

The National Weather Service's Storm Prediction Center is the best place to research the history of severe weather patterns back to the year 1950. By providing weather forecasts to local field offices across the nation, the Storm Prediction Center's goal is education and quick communication to representatives from each state. In addition to forecasting hurricanes and tornadoes, future plans include researching other weather events such as high winds, dense fog conditions, and air pollution. The around-the-clock operation provides important information to the public, too, all of which may be viewed on the center's Web site at http://www.spc.noaa.gov

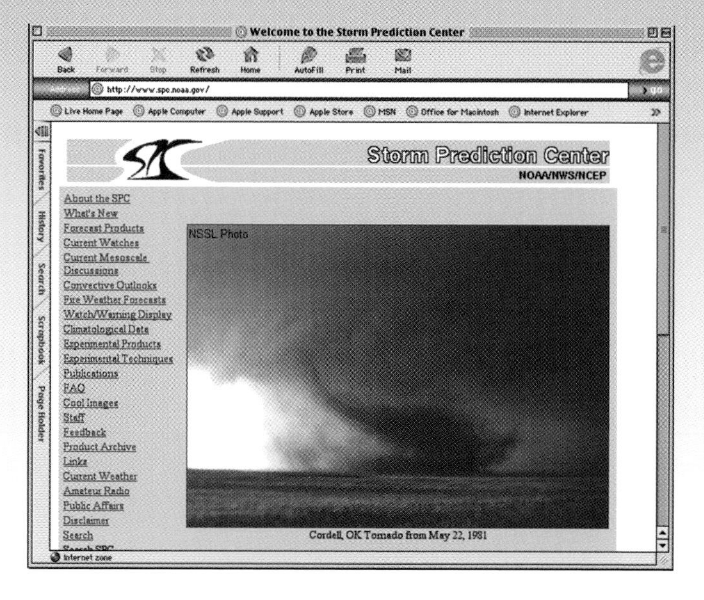

and journalists who do spend their time following severe storms all over the globe, but even they experience times when there are no storms to chase.

Because tornadoes and hurricanes are very different from each other, the ways in which they are chased are not the same. On one hand, tornadoes are comparatively small and begin and end on land, a fact that makes them accessible to everyone, whether hobbyist or professional meteorologist. On the other hand, hurricanes begin at sea and, because they are so massive, are first observed by satellite. Also, hurricanes are more predictable than tornadoes, so coastal areas in the line of danger are often evacuated, eliminating casual hurricane chasers. As for flying into hurricanes, only a handful of highly trained professionals ever take such a trip.

Still, depending on what your interests are or where your skills lay, there are a variety of careers in which you could spend part of your hours on the job chasing the actions of severe storms. Be forewarned, though: While chasing after some of the scariest weather on Earth may make perfect sense to you, don't be surprised if your friends and family don't quite understand your interest in something they may consider very unusual.

Gregg Potter, lead meteorologist for F5 Torando Chasing Safaris, gets paid to show clients tornadoes firsthand. Here he unloads equipment to film a potential tornado north of Wichita, Kansas, in June 2001.

Tornado Research Meteorologist

Even with computers and other types of technical equipment, we still have many unanswered questions about tornadoes. It is a research meteorologist's job to not only search for answers, but also to come up with new questions. During tornado season, most researchers spend at least some time in the field chasing tornadoes. While they are there, it is their job to witness as many storms as possible. While doing so, they gather data from a variety of measurement devices such as a radiosonde, which is a package that is carried into the sky like a weather balloon. Researchers input information like air temperature, air pressure, humidity, and more into laptop computers that graph and tabulate the raw data.

Daphne Zaras, who often spends her days in the field following tornadoes, explains a typical experience: "When we head out to the field, there are usually two or three of us to a car, with anywhere from three to twelve cars. On a typical day, we're on call for twelve hours at a stretch. The most dangerous part of storm chasing is driving. This is because when other people on the road see storms, they are often shocked, and sometimes they'll stop in the middle of the road to look at the approaching storm. That's why only one team member drives. Every car also has a team leader who collects data and there are often one or two other people who are in charge of navigating, as well as shooting videos and photographs. If we're using a new computer program, we will bring along a programmer who can help us work through any problems we might have using it for the first time. Much of the time out there is just down time, just waiting between storms. Am I scared of tornadoes? I used to be. They're very powerful. But we never approach a tornado without an option for a quick getaway. And many times, we're so busy working that we don't have time to feel afraid. I'm still scared of night tornadoes though. You can't see those coming."

When the season ends, researchers head back into the lab until the start of the next season, sorting through the information they collected and crunching numbers in an attempt to further understand what makes tornadoes tick—or rather, twirl.

A person who is thinking about becoming a tornado research meteorologist should definitely enjoy math and science as much as they do chasing tornadoes!

Numbers are a big part of the job. The best preparation for someone considering this type of career is a lot of coursework in math, physics, and chemistry. Having a good understanding of how computers work—including how they are programmed—would also be helpful.

Hurricane Research Meteorologist

Like tornado researchers, hurricane research meteorologists seek answers to the many puzzles we have about

Meteorologist Joseph Cione holds an airborne expendable bathythermograph (AXBT), which measures ocean temperatures that affect how large storms may become.

hurricanes. They work especially hard at improving their ability to forecast hurricanes and their weaker counterparts, tropical storms, partially because issuing warnings is very expensive: tourism drops, shopping declines, and residents are forced to spend money on safeguarding their homes and businesses. Outside of a major city, the average cost of issuing a hurricane warning ranges from $500,000 to $1 million per mile.

When researchers do hunt hurricanes (again, only for a few months out of each year), they use planes to fly right through them. Throughout hurricane season, researchers at the Atlantic Oceanographic and Meteorological Laboratory fly into hurricanes along the Atlantic coast to collect data they will study for the rest of the year in their labs.

During storm season, the U.S. military also flies into storms to collect data, which they provide to the National Hurricane Center. The six-member crew includes two pilots, a navigator, a flight engineer, a mission director/weather officer, and the operator of the dropsonde, an instrument used to measure hurricane winds and pressure. Nicknamed the Hurricane Hunters, they are all members of the U.S. Air Force Reserve. For someone who would like to someday be a member of the Hurricane Hunters, in addition to having a strong

To collect data, military pilots like Captain Troy Anderson, pictured here flying through Hurricane Luis in 1985, are dispatched into the hearts of storms.

background in math and science, he or she would also have to join the U.S. military.

Tornado Expedition Guide

Anyone who has ever had a hobby knows that in no time at all, it can go from being a part-time pursuit to a full-blown obsession. This is also true for tornado expedition guides. They may begin by chasing an

Brad Colman (*far left*) of the National Weather Service, and Brian Colley, a professor at SUNY Stony Brook in New York, discuss a precipitation system.

occasional storm and suddenly find themselves spending every spare minute racing back and forth on country roads in hot pursuit of twisters. Eventually, they gather enough experience and expertise to start a company. For obvious reasons, tornado expedition companies are almost always located in regions of the country where tornadoes are extremely common. Martin Lisius's business, for example, is located in Arlington, Texas.

If you are leading people on a chase of something as dangerous and unpredictable as a tornado, there is much more involved than just listening to weather reports and jumping in the car and driving. Tornadoes are hard to find! In order to increase their chances of finding one, tornado expedition guides spend many hours studying weather forecasting techniques and learning as much as they can about the severe storms that produce tornadoes.

Lisius explains, "Tornadoes are one of the rarest types of severe weather. They originate from super cell thunderstorms, and that type of thunderstorm is uncommon, too. If the super cell does produce a tornado, then there's only a two percent chance that it will be an F4 or F5. If you're good, and you chase a lot, you see an F5 on occasion, but not very often."

In addition to sharing their knowledge of storms with tour members, most guides feel it is important to alert local weather broadcasters to approaching tornadoes and share their discoveries with researchers.

For anyone who wants to be a tornado expedition guide, having a responsible personality is just as important as having an interest in severe weather. On every chase, guides risk exposing their expedition members to

Storm chaser Martin Lisius uses various communications equipment in his van to track storms for his storm expedition company, Tempest Tours, Inc.

lightning, hail, and the tornado itself, not to mention encountering the aftermath of a tornado, which may include people who are injured and destroyed property.

Weather Forecaster

Students who study meteorology don't always become research scientists—sometimes, they choose to become weather forecasters. When you think of a weather forecaster, the first person that usually comes to mind is a television personality—a person who, for example, lets you know whether or not your baseball game will be cancelled because of rain or if the temperatures are warm enough for outdoor swimming. Weather forecasters, however, also predict weather for the U.S. government, military, and even private companies, including television and radio stations.

The job of a weather forecaster is to interpret incoming data from a variety of measurement devices and provide accurate weather predictions, along with warnings and safety instructions. Traditionally, forecasters have not chased storms, but as the public has

Mike Kotyk, a weather forecaster with the U.S. Navy, tracks the progress of Hurricane Allison along the East Coast in 2001.

grown increasingly fascinated by severe weather, some television stations either assign reporters to cover storms in the field or hire temporary storm chasers as field reporters. This is particularly the case with tornadoes. In the Midwest, where tornadoes are most common, local television stations will sometimes interrupt programming to show live broadcasts of severe storms in progress.

If the idea of becoming a weather forecaster appeals to you, then you will need to do well in math and science. Most weather forecasters also have a degree in meteorology. If you plan to work for a television or radio station, then you should be as comfortable with the thought of speaking in front of large groups of people as you are with seeing a severe storm on the horizon.

Severe Storm Journalist, Photographer, and Videographer

Given the spectacular nature and rarity of both tornadoes and hurricanes, it is not surprising that

Warren Faidley, who chases and photographs storms full-time, holds up a storm photo during a Las Vegas convention on severe weather.

photographs and videos of them are always in high demand. As a result, storm chasers often supplement their hobby with income earned from selling photographs and videos of severe storms they have chased. These images are purchased by local news channels, national television stations, magazines, movie studios, and even advertisers. Images and videos shot by storm chasers have been featured in the movie *Twister*, the weekly show *Buffy the Vampire Slayer*, and a variety of music videos on MTV. Many chasers also sell their videos and photographs on the Internet to interested individuals.

In addition to photographing and videotaping storms, tornado and hurricane hunters also write about them. As journalists covering violent storms, they might report about the storms for local newspapers or contribute information on a regular basis to publications whose focus is severe storms. (In a bit of a twist, it has also become common in recent years for magazines and television news shows to investigate and report on the lives of storm chasers!)

Though most storm chasers write about and photograph storms on a part-time basis, there is the occasional individual who follows storms all over the world. In these isolated examples, he or she is able to do so on a full-time basis. Warren Faidley, for example, is a full-time photojournalist who specializes in shooting severe storms. His images have been seen in books, advertisements, magazines, and public safety publications. He was also the initial consultant for the film *Twister*. Having witnessed both tornadoes and hurricanes, he does consider one more dangerous than the other.

He commented, "Hurricanes are likely the most dangerous storms to chase. Instead of shooting the storm from the outside, like tornadoes and lightning, I'm working inside the storm. You cannot simply run away. Once you're committed to a location, you better hope it survives!"

6 On the Path

Do you envision yourself one day chasing down a tornado? Or, like the Hurricane Hunters, flying right through the eye of a hurricane? If so, the best way to prepare for such a career is to focus on subjects such as mathematics and science, including advanced math and science classes. Since so much research is done with the aid of computers, having a good understanding of basic computer programming would also be a good idea.

Believe it or not, you can also start "chasing" severe storms as a hobby from the safety of your home. Many people who now chase storms professionally started out in their teens as certified storm spotters. In communities where severe storms are

Bob Glancy of the NWS (National Weather Service) shows weather disaster images during a class in Weld County, Colorado, the U.S. county with the most reported tornadoes.

common, many organizations sponsor training sessions that can lead to certification. The job of a spotter is to watch the sky for possible storms, which he or she reports on amateur radio frequencies. (Amateur radio comprises a community of people who use radio transmitters and receivers to communicate with other amateur radio operators.) Storm spotters are often police officers and emergency workers, but they are additionally ordinary people who just want to help or have a unique interest in severe weather events. They are valuable assets to the community and perform an important public service that helps to insure the safety of residents and their property.

On the Internet, there are several Web sites that offer up-to-the-minute information about storm activity across the United States. Visitors to the Federal Emergency Management Agency's (FEMA) site (http://www.fema.gov/kids) can download maps to track the movement of hurricanes.

It might be easy to forget just how unpredictable and dangerous severe storms can be. Sure, people have flown right through hurricanes and stood a stone's throw away from twisters without harming themselves, but one thing they never forget is just how dangerous those situations could have been. Just like wild animals, severe storms, however beautiful, must be respected. After all, Mother Nature will always remind people just how unpredictable and strong she can be. For people who respect nature, it is just as likely to provide experiences that are indescribably beautiful and unforgettable.

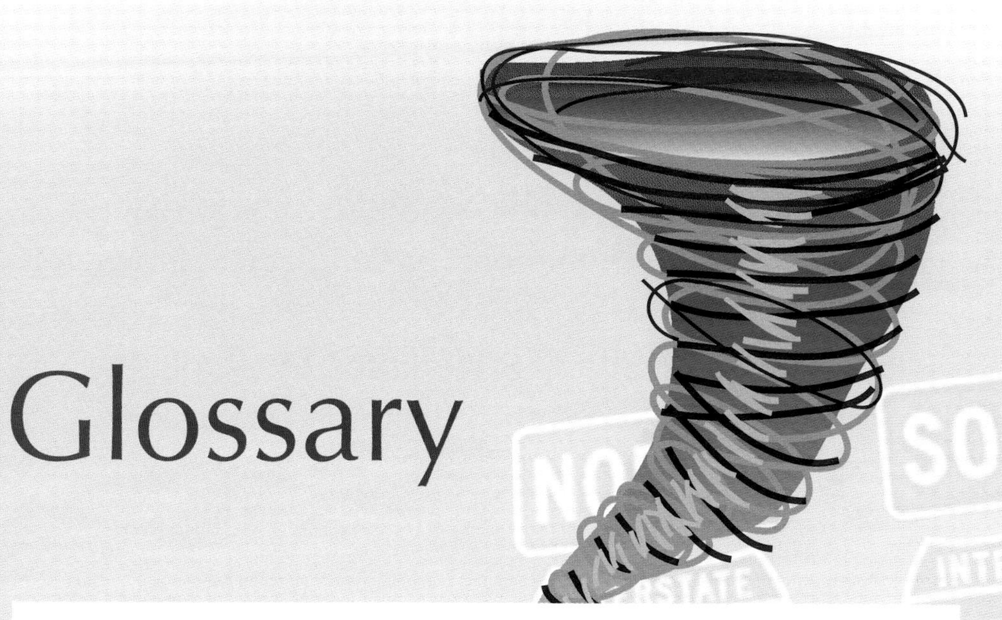

Glossary

amateur One who engages in study, pursuit, or sport as a pastime rather than a profession.

condensation A liquid formed by condensed water.

dropsonde A radiosonde (see definition) dropped by parachute from a high-flying airplane.

eye A region in the center of a hurricane where the winds are light and skies are clear to partly cloudy.

eyewall A thick wall of clouds that surrounds the eye of a hurricane.

flora Plant life, especially those characteristic of a certain region or environment.

Fujita-Pearson Tornado Scale A scale ranging from zero to five based on a tornado's intensity. It can be used to estimate the potential property damage expected from a tornado.

genesis stage The first phase of hurricane formation.

hail Precipitation in the form of compacted snow and ice particles.

Hurricane Hunters A U.S. Air Force squadron that flies into hurricanes to collect meteorological information on them.

intensification stage The second phase of hurricane formation, in which the hurricane becomes stronger.

meteorologist A person who studies the weather and weather conditions.

meteorology The science dealing with the study of the atmosphere, especially weather and weather conditions.

midget A very small hurricane.

phenomena Rare or significant facts or events.

radar A method of detecting distant objects and determining their position, velocity, or other characteristics by analysis of high-frequency radio rays reflected from their surface.

radiosonde A weather measurement device like a weather balloon that is released into the sky to measure air temperature, pressure, and humidity.

Saffir-Simpson Hurricane Scale A scale ranging from one to five based on a hurricane's intensity. It can be used to estimate the potential property damage and flooding expected along a coast from a hurricane.

satellite A man-made object or vehicle intended to orbit Earth, the moon, or another celestial body, which sometimes gathers data (including photographs) about the atmosphere.

super cell thunderstorm A thunderstorm that exhibits certain well-defined characteristics and usually produces severe weather.

typhoon A hurricane that forms over the western Pacific Ocean.

For More Information

Atlantic Oceanographic and Meteorological Laboratory (AOML)
4301 Rickenbacker Causeway
Miami, FL 33149
(305) 361-4450
Web site: http://www.aoml.noaa.gov

Digital Cyclone/EarthWatch Weather on Demand
5125 County Road 101, Suite 300
Minnetonka, MN 55345
(952) 974-3300
Web site: http://www.earthwatch.com

Federal Emergency Management Agency (FEMA)
500 C Street SW

Washington, DC 20472
(202) 646-4600
Web site: http://www.fema.gov

Hurricane Hunters
United State Air Force Reserve
403rd Wing
701 Fisher Street
Keesler AFB, MS 39534
(228) 377-2056
Web site: http://www.hurricanehunters.com

National Oceanic and Atmospheric
Administration (NOAA)
14th Street and Constitution Avenue NW, Room 6013
Washington, DC 20230
(202) 482-6090
Web site: http://www.noaa.gov

National Severe Storms Laboratory (NSSL)
1313 Halley Circle
Norman, OK 73069
(405) 360-3620
Web site: http://www.nssl.noaa.gov

The Tornado Project
P.O. Box 302
St. Johnsbury, VT 05819
Web site: http://www.tornadoproject.com

Weatherwise magazine
1319 18th Street, NW
Washington, DC 20036-1802
(202) 296-6267
Web site: http://www.weatherwise.org

Web Sites

Due to the changing nature of Internet links, the Rosen Publishing Group, Inc., has developed an online list of Web sites related to the subject of this book. This site is updated regularly. Please use this link to access the list:

http://www.rosenlinks.com/eca/hhtc/

For Further Reading

Allaby, Michael. *DK Guide to Weather*. New York: DK Publishing, 2000.

Challoner, Keay. *Eyewitness: Hurricane and Tornado.* New York: DK Publishing, 2000.

Davidson, Keay. *Twister.* New York: Pocket Books, 1996.

Davies, Pete. *Inside the Hurricane.* New York: Henry Holt and Company, 2000.

Stein, Paul. *Storms of the Future* (Library of Future Weather and Climate). New York: The Rosen Publishing Group, Inc., 2001.

Toomey, David. *Stormchasers: The Hurricane Hunters and Their Fateful Flight Into Hurricane Janet.* New York: W.W. Norton & Company, 2002.

Trueit, Trudi. *Storm Chasers.* New York: Franklin Watts, Inc., 2002.

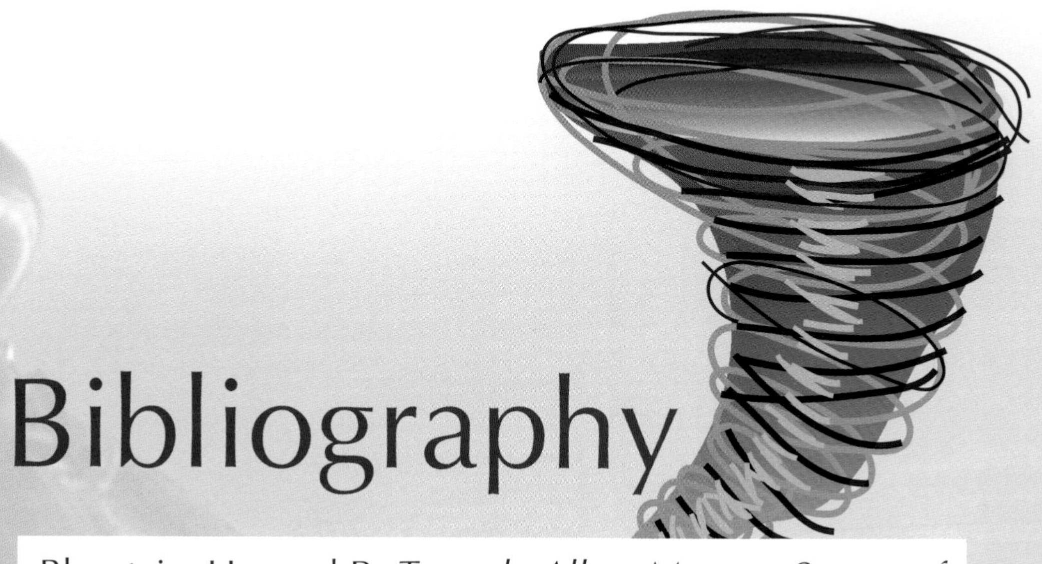

Bibliography

Bluestein, Howard B. *Tornado Alley: Monster Storms of the Great Plains.* New York: Oxford University Press, 1999

England, Gary A. *Weathering the Storm.* Oklahoma City, OK: University of Oklahoma Press, 1996.

Fitzpatrick, Patrick. *Natural Disasters: Hurricanes.* Santa Barbara, CA: ABC-CLIO, 1999.

Grazulis, Thomas P. *The Tornado: Nature's Ultimate Windstorm.* Oklahoma City, OK: University of Oklahoma Press, 2001.

Longshore, David. *Encyclopedia of Hurricanes, Typhoons, and Cyclones.* New York: Facts on File, 2000.

Mosier, Tim B. *Twisters in the Heartland.* Orlando, FL: Rivercross Publishers, 1998

Reiss, Bob. *The Coming Storm: Extreme Weather and Our Terrifying Future*. New York: Hyperion, 2001.

Renner, Jeff. *Lightning Strikes: Staying Safe Under Stormy Skies*. Branson, MO: Mountaineer Books, 2002.

Rosenfeld, Jeffrey. *Eye of the Storm: Inside the World's Deadliest Hurricanes, Tornadoes, and Blizzards*. New York: Penum Trade, 1999.

Rubin, Louis D. *The Weather Wizard's Cloud Book: How You Can Forecast the Weather Accurately and Easily by Reading the Clouds*. Chapel Hill, NC: Algonquin Books, 1989.

Sheets, Bob, and Jack Williams. *Hurricane Watch: Forecasting the Deadliest Storms on Earth*. New York: Vintage Books, 2001.

Williams, Jack. *The Weather Book*. 2nd ed. New York: Vintage Books, 1997.

Index

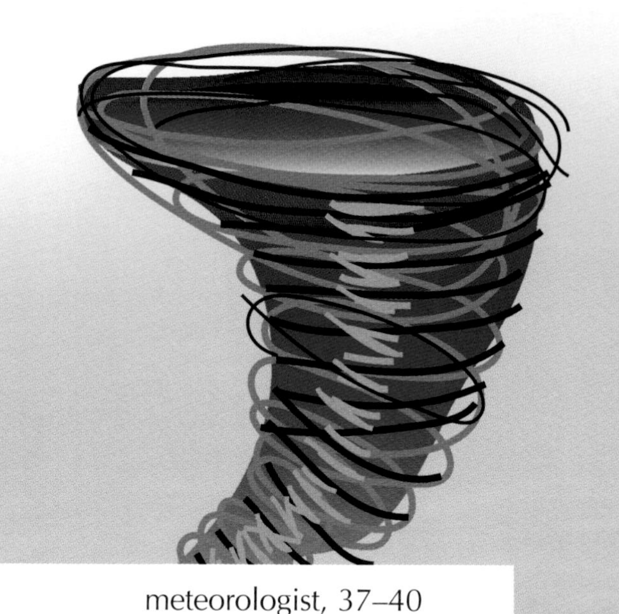

About the Author

Lois Sakany is a writer who lives in Brooklyn, New York. This is her first book for young readers, and although she's never chased a hurricane or a tornado, she would like to someday.

Photo Credits

Cover © Howard Bluestein/Photo Researchers; pp. 7, 10 © WeatherStock; p. 13 © Museum of the City of New York/Corbis; p. 15 © Bettmann/Corbis; pp. 19, 25, 34, 37, 39, 40, 42, 44, 46, 50 © AP/Wide World Photos; p. 22 © The Everett Collection; p. 26 © Historic NWS Collection; p. 29 © Raymond Gehman/Corbis, p. 32 courtesy of NOAA/NWS.

Design and Layout

Les Kanturek